Asset Managers

Joy Yongo

Series Editor **Casey Malarcher**

Level 1 - 5

Asset Managers

Joy Yongo

Series Editor: Casey Malarcher
Content Editor: Anne Taylor
Copy Editor: Liana Robinson
Cover/Interior Design: Highline Studio

ISBN: 978-1-943980-37-6

10 9 8 7 6 5 4 3 2 1
22 21 20 19 18

Photo Credits

Contents

What Is Investing?

One way to make money is to invest it.
People who invest are called investors.
They put money away for a while.
Their hope is to make more money in the future.

◀ Making money grow

Stock market ▶
prices

There are different ways to invest.
The first is to buy stocks.
When a person buys a stock, he or
she buys a small part of a company.
When the company makes money,
so do the people who hold stocks.

BUY ◀ Buying stocks

Another way to invest is to buy bonds.
Buying bonds is like letting another person or company borrow your money.

Borrowing money ▶

The company that takes the bond pays back the money a little bit at a time.

It must pay interest or extra money.

This is how you make money from others who borrow from you.

The interest a person pays ▶ adds up over time.

Stocks and bonds are called assets.

These are what investors buy and sell to make money.

Different kinds of assets

The Job of an Asset Manager

Sometimes, people or companies want to invest.

Maybe they have a lot of money to invest.

Maybe they have only a little money to invest.

◀ Money to invest

A house as an investment goal

People have different investment goals.
A goal could be buying a house.
It could be saving for a child's
university study.
It could be saving for the time after
a person stops working.

Planning for ▶
the future

Some people may want to invest, but they don't know how.

They also do not have the time to study about investing.

What can they do?

They can get help from an asset manager.

Some people don't ▶
know how to start
investing.

Asset managers help people meet their investment goals.

The asset manager looks for which assets are best to reach the investor's goals.

They look for assets that help their clients make money.

◀ An asset manager researching online

An asset manager talking to clients

The asset manager tells the client which assets are good to buy.

When the client buys an asset, it goes into an investment portfolio.

Watching for the best time to buy or sell

A portfolio includes all of the stocks and bonds that the investor has.

The asset manager takes care of the client's portfolio.

He or she buys and sells assets for the client.

STOCK MARKET

◀ A bear market has falling prices. This is when investors sell more. A bull market has rising prices. This is when investors buy more.

Asset managers watch how assets change.

For example, stock prices go up and down every day.

An investor will want to buy a stock at a cheap price.

Prices can rise and fall quickly like a rollercoaster. ▶

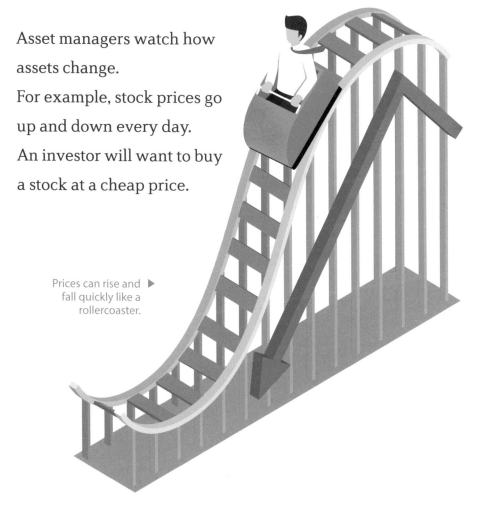

◀ Selling stocks

When the stock price is high, the investor can sell it.

This is another way investors make money from stocks.

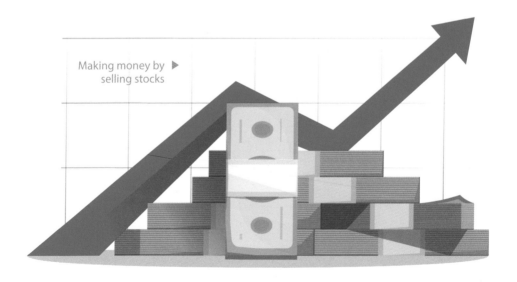

Making money by ▶
selling stocks

Asset managers will watch the stock prices.

They will make reports for the investor to see.

Creating a report for an investor

A client looking at his ▶
investments

◀ Stocks can have
high or low risk.

The asset manager will also look to see if an investor
may lose money.

Then, he or she will help the investor decide what to
do with the asset.

How to Be an Asset Manager

A university degree is not needed to be an asset manager.
But a degree makes it easier to get a job as an asset manager.
Asset managers usually study finance or business in university.

Students studying

University classes help students learn how to take care of investment portfolios.

But there is more to know than just that.

At a company, a new asset manager may work with someone more experienced in investing.

Some companies may also give more training.

Working with a more experienced asset manager at a company

Asset managers should have many skills.

They should be good with numbers.

They should also be good with details.

They are working with other people's money!

◀ Working with people's money

A calculator

Asset managers should know how to ask the right questions and find the answers.

Knowing how to solve ▶
problems

Asset managers need to make reports.

So they should know how to use computer programs.

They should also keep things in order.

They need to know where to find things quickly and easily.

◀ Keeping things organized as an asset manager

Some asset managers work by themselves.

Some asset managers work in teams.

◀ A team of asset managers

Working alone

Listening to clients

But all asset managers should work well with people.
They need to listen to what their clients want and need.
They want to keep their clients happy.

Looking to the Future

People will continue to invest.

Companies will continue to invest.

Both will need people who know how to invest well.

Investing well to make money grow

A career as an asset manager

Do you like working with numbers?

Can you take care of someone's money?

Think about becoming an asset manager!

Comprehension Questions

1. Which is a type of investment?
   (a) Stocks
   (b) Bonds
   (c) Both a and b
   (d) None of the above

2. People or companies can begin investing when . . .
   (a) they take an investing test.
   (b) the market offers bonds for new investors.
   (c) asset managers open their accounts.
   (d) they have some money.

3. Which is an example of an investment goal?
   (a) Buying a house
   (b) Paying for college
   (c) Saving for the future
   (d) All of the above

4. An asset manager . . .
   (a) takes care of people's financial portfolios.
   (b) takes people's money.
   (c) creates stocks.
   (d) teaches students about money.

5. If you want to make money, when should you sell a stock?
   (a) When the price is low.
   (b) When the price is high.
   (c) When the price is the same.
   (d) When you don't know the price.

Glossary

- **asset** (n.) a thing of value, especially property, that a person or company owns, which can be used or sold to pay debts

- **bond** (n.) an official document that says you will be paid a certain amount of money because you have lent money to a government or company

- **borrow** (v.) to take and use something that belongs to somebody else, and return it to them at a later time

- **client** (n.) a person who pays for services or advice

- **degree** (n.) a qualification obtained by students who successfully complete a university or college course

- **goal** (n.) something that you hope to achieve

- **invest** (v.) to buy property or shares in a company in the hope of making a profit

- **manager** (n.) a person whose job is to lead employees or maintain programs

- **portfolio** (n.) all of the assets owned by a particular person or organization or a list of all the owned assets

- **stock** (n.) a share in a company that people buy or sell when investing

Notes

Here are some other types of assets and investments. Readers may enjoy researching these topics to learn more about things that asset managers work with.

Annuity: A financial product which accepts money from an individual and grows as a fund. The person who receives the annuity receives a fixed amount of money every year for the rest of his or her life.

Certificate of Deposit (CD): This is considered to be the safest form of investment. An individual agrees to lock up some money in a bank so that he or she can receive a higher interest rate on the money that is saved.

Investment Fund: Investing money as a group instead of as an individual. Each investor in the group receives the benefits of the group investment.

Mutual Fund: An investment fund managed by a professional. It uses money from many investors to buy different types of stocks, bonds, and other assets.

Exchange-Traded Fund (ETF): A type of investment fund that is like a mutual fund. However, ETF prices go up and down daily, and such funds can be bought and sold like stocks.

List of Books

LEVEL 1

1. Robotics Engineers
2. Cyber Security Experts
3. Medical Scientists
4. Social Media Managers
5. Asset Managers

LEVEL 2

1. Drone Pilots
2. App Developers
3. Wearable Technology Creators
4. Computer Intelligence Engineers
5. Digital Modelers

LEVEL 3

1. IoT Marketing Specialists
2. Space Pilots
3. Water Harvesters
4. Genetic Counselors
5. Data Miners

LEVEL 4

1. Database Administrators
2. Nanotechnology Research Scientists
3. Quantum Computer Scientists
4. Agricultural Engineers
5. Intellectual Property Lawyers

"The future of the economy is in STEM. That's where the jobs of tomorrow will be."

James Brown (Executive Director of the STEM Education Coalition in Washington, D.C.)

Data from the US Bureau of Labor Statistics (BLS) support that assertion. Employment in occupations related to STEM—science, technology, engineering, and mathematics—is projected to grow to more than 9 million by 2022 [in the US alone] … Overall, STEM occupations are projected to grow faster than the average for all occupations.

from *STEM 101: Intro to Tomorrow's Jobs* **US Bureau of Labor Statistics**